Of Ticks and Islands

Memoirs from the Lyme & Vector-borne Disease Laboratory of the Maine Medical Center Research Institute

by Peter W. Rand, MD

The material in this book is intended for education.

ISBN: 978-0-9990454-1-1

Ebook ISBN: 978-0-9990454-0-4

CONTENTS

FOREWORD

New infectious diseases represent a significant threat to human health and are emerging worldwide at an escalating pace. Maine and northern New England are not immune from this alarming trend. Most infectious pathogens arise as a result of changes in our natural environment, whether due to climate change, ecological disruptions, or the transport of invasive species of ticks or mosquitos that can transmit disease to a new geographic area. The critical task of understanding, tracking and responding to these new diseases, requires that researchers combine work in the field, the laboratory, and the clinic. The Lyme & Vector-borne Disease Laboratory (VDL) at Maine Medical Center Research Institute (MMCRI) is one of the entities in New England that integrates these approaches.

Founded in 1990, the VDL has played a major role in research on the causes of new infectious diseases in northern New England and on innovative strategies for preventing them. What started as a small project focusing on the spread of a newly arrived tick species transmitting Lyme disease grew over the subsequent three decades into a laboratory devoted to research on seven new diseases, all born by ticks and mosquitoes, that put Mainers at risk in their own backyards.

The work that Peter Rand describes in this memoir illustrates the VDL's work on Lyme disease, and it represents the cornerstone of the VDL's approach to its investigation of several disease threats. In over 60 peer reviewed research articles, the lab has made numerous contributions to the literature on emerging infectious agents: it has delineated the environmental risks that lead to the establishment of tick populations in Maine; it demonstrated the prevalence of newly discovered pathogens in Maine ticks; and it devised ecologically sensitive strategies for controlling them, including testing of botanical agents. In the process, the lab has trained young researchers who have gone on to devote their careers to these serious challenges and have themselves trained the next generations of medical students, residents, and faculty members. The lab also has created a tradition of community outreach to educate the public regarding the risks and management of tick-born diseases, with programs stretching from Acadia to Kittery and from Portland to Waterville.

What Dr. Rand does not mention in this memoir is that he was also instrumental in early research activities at Maine Medical Center that resulted in the founding of MMCRI in 1991. As one of the first dedicated—and NIH supported—research scientists at Maine Medical Center, he pioneered work on blood viscosity that enabled effective use of the heart-lung machine. Later, he turned his attention to environmental health risks such as paralytic shellfish poisoning and radon exposure in homes. When Lyme disease arrived in Maine, Dr. Rand was ready to apply his years of research experience in and out of the lab to an issue that required innovative approaches.

Innovation is a hallmark of research at MMCRI. Our work in molecular biology has contributed crucial understanding of the cellular processes across a range of significant diseases, and our health outcomes research group has provided keen insights and tools for improving population health in the state of Maine and beyond. All of this work has led to discoveries that can improve, extend, and save lives within our own hospital and state-wide health network, but also far beyond. Furthermore, our research activities foster an environment of innovation and inquiry that informs the commitment to medical education and connects us to leading research and health care institutions around the world. In a way, Dr. Rand has written *Of Ticks and Islands* for all of us, capturing our personal commitments to innovative research that changes the world.

Donald L. St. Germain, M.D., Vice President, Research, Maine Medical Center; Director, Maine Medical Center Research Institute

Robert P. Smith, Jr., M.D., Director, Lyme and Vector-borne Disease Laboratory, Maine Medical Center Research Institute; Professor of Medicine, Tufts University School of Medicine

PREFACE

This booklet is not a scientific treatise. It is a memoir of some unique field research carried out by a small team of investigators bent on helping out some islanders faced with an increasing threat to their health and well-being. It recounts the sometimes colorful interplay among investigators, townspeople, and participants from both state and federal agencies. It has a happy ending.

My colleagues and I comprise the Lyme & Vector-borne Disease Lab at the Maine Medical Center Research Laboratory in Scarborough, Maine. We've been studying the ecology of tick- and mosquito-borne diseases since the first tick vectors of Lyme disease came over the New Hampshire border in the 1980s. Prior to that time, my research associate, Eleanor Lacombe, and I had been focusing on environmental health issues since the 1970s, the last bit studying the frequently alarming levels of radon gas in domestic water supplies and homes in southern Maine. Persuading people to let us place detectors in their cluttered cellars or in their toilet tanks was our baptism in field epidemiology.

Back in the late 1940s, I was extremely fortunate to sail with explorer Donald B. MacMillan on three trips to the Arctic in the Schooner *Bowdoin*. From that, and from the seven subsequent decades my wife and I cruised the Maine coast, I developed the habit of keeping more or less legible trip logs. Were it not for that habit, those exposures to the natural world, and serendipitous events, this booklet would not have come to pass.

On my first trip of a lifetime to coastal Maine, they said I fit into a bureau drawer. Twenty-four years later I married a native. Despite my father's admonition not to let geography dictate my career, in 1955 I moved directly from medical school to the Maine Medical Center (then Maine General Hospital). Post-graduate training in cardiology involved me directly in the care and operative support (perfusion) of the hospital's first open-heart surgery patients. The program was very early nationally, incredibly brave, thankfully successful, but like all the rest, essentially experimental. With each patient issues were encountered that needed research.

So it was that in 1960, a grant was submitted to, and awarded by, the National Institutes of Health which allowed us to establish the hospital's Research Department. My role, for the first decade, was to conduct studies related to open-heart surgery and the flow properties of blood. But then our research shifted to focus on regional environmental health issues, eventually to the Environmental Protection Agency-supported radon study mentioned above. In December, 1987, the agency's leadership changed, and with it, precipitately, their research priorities. We were left hanging in the breeze.

The story picks up from there.

CHAPTER 1.

MONHEGAN

Two events which took place over a half-century ago on an isolated island off the coast of Maine set the stage for this story about ticks in Maine. Details of the first event, which occurred in 1954, have vanished from the islanders' memories, although an impression of what happened was published 40 years later in a Portland newspaper:

> The first white-tailed deer arrived on Monhegan Island the same way thousands of camera-toting and L. L. Bean bag-carrying visitors show up each summer – by ferry from Port Clyde. State wildlife officials, at the request of islanders, imported nine of the animals from shore to amuse tourists and give locals game to shoot. It was the end of the Eisenhower administration. For years the deer were scarcely seen. But as the herd grew, the deer began to stray from the quiet of Cathedral Woods into the meadows in and around the fishing village, searching for food.
>
> *Maine Sunday Telegram*, April 12, 1998.

Maybe the number of deer was not nine. In a letter dated April 12, 1954 to an inquisitive islander, the Chief of the Game Division of the then-Maine Fish and Game Department writes that he honored the request made at the island's March Town Meeting by releasing six deer, trapped on Swan Island in the Kennebec, on Monhegan. Some islanders say they always heard the deer came over by lobster boat, which might have afforded a much more dramatic opening to what follows, but such a trip has yet to be verified.

Six years later, in a seemingly unrelated project, mammalogist Garrett Clough, husband of a respected Monhegan artist, conducted small mammal trapping surveys on several midcoast islands. On Monhegan he found no mice, voles, chipmunks, squirrels, or other small mammals. None has been reported since.

The dominant event for this story, however, took place three states away in 1975, when Allen Steere, a physician from Yale University, was called at the urging of several mothers to evaluate 12 children in Lyme, Connecticut who had come down with fevers and painful, swollen joints. Their illnesses were initially attributed to juvenile rheumatoid arthritis, but their symptoms weren't typical, nor does that condition occur in geographic or temporal clusters. Common to many was a history of a recent tick bite, and some had developed an expanding ring-like rash at the site. Steere described this tick-borne illness as a new disease which he named after the town where he first recognized it. The cause of Lyme disease was not identified, however, until 1982, when Willy Burgdorfer, a microbiologist working in a federal lab in Montana, saw coiled bacteria – spirochetes – in the gut tissues of deer ticks sent to him from Shelter Island, New York. These were subsequently shown to be the cause of Lyme disease. In his honor, this new pathogen was named *Borrelia burgdorferi.*

The deer tick itself was recognized as a new species by Dr. Andrew Spielman, an internationally respected medical entomologist at the Harvard School of Public Health. He named it *Ixodes dammini* after a close friend, noted Harvard pathologist Gustav Dammin.

Public health entomologist Andrew (Andy) Spielman, ScD (1930-2006). Photo used with permission from Harvard School of Public Health.

Andy demonstrated that his tick transmitted both the Lyme pathogen as well as that of a malaria-like illness, babesiosis, which had plagued the residents of islands off Cape Cod beginning in the 1960s. He showed that *I. dammini* larvae and nymphs fed primarily on white-footed mice, which, as natural reservoirs of the Lyme and babesia agents, both nourished and infected the ticks in the process. The larvae, after overwintering and molting to nymphs, feed and infect the next year's crop of reservoir mice and other small mammals. These nymphs, feeding in the late spring and early summer, are also the primary transmitters of Lyme and other tick-borne diseases in humans and their domestic animals. Fed nymphs then molt to male and female adults which primarily seek white-tailed deer, males to mate and females to obtain the blood meal that will nourish the development of eggs. Up to 3000 eggs per female are deposited in late May and June, continuing the life cycle of this major disease vector.

So while mice are the ultimate source of infection, white-tailed deer are primarily responsible for maintaining the tick's life cycle, a proposition to be explored in the following pages. This project, indeed all of the early work on tick-borne diseases on the islands of Maine, benefited hugely by having Andy Spielman as our mentor and friend.

In the 1990s, because some of the entomologist hierarchy were not convinced that Andy's *I. dammini* was sufficiently different from a similar southern tick called *I. scapularis*, a taxonomic war ensued. Although these species names are now recognized as synonymous, *I. scapularis* has prevailed in the literature. For its official non-scientific name, entomologists prefer the term "black-legged tick," but given the strong preference this species has for whitetails, it is commonly known as the "deer tick," as it will be referred to hereafter.

In the late 19th century and throughout the 20th, conditions in the northeastern United States became progressively favorable for the reproduction and spread of deer ticks. As more and more farms were abandoned during the Industrial Revolution, their open fields gave way to new growth, providing food and cover for white-tailed deer and shade and moisture for ticks. Later, as affluence allowed city dwellers to move to the country where they frequently posted their properties with "No Hunting" signs, the deer thrived and deer ticks multiplied.

Songbirds are also hosts for deer ticks – larvae and nymphs but not adults – and several species may harbor Lyme bacteria with which they infect deer ticks feeding on them. In southern New England, where deer numbers had increased dramatically, abundant overwintered larvae and nymphs were available to attach to spring-migrating songbirds. Having fed during transit, these replete – and now infected – ticks might detach hundreds of miles to the north.

Prior to the 1980s, deer ticks and tick-borne diseases were rarely seen north of Cape Cod. In 1982 however, a focal outbreak of Lyme disease occurred near coastal Ipswich, Massachusetts, just 25 miles south of the Maine border, where 35 percent of residents living within three miles of a game preserve contracted Lyme disease. Shortly after, a taxidermist in Connecticut found deer ticks on the hide of a deer shot in Jackman, Maine. Then, in 1987, entomologist Howard Ginsberg, working for Acadia National Park, collected 17 deer tick nymphs in Bar Harbor, two of which were infected with *Borrelia burgdorferi*. The Lyme zoonosis had come to the Pine Tree State.

Fortuitously, later that year, two investigators at the Research Department of the Maine Medical Center working on radon and environmental health had their expectations for continued EPA funding dashed by a sudden shift in the agency's priorities. That made them available when newly-arrived infectious disease specialist Dr. Robert Smith asked if the emergence of Lyme disease in the state might not be an environmental health issue worth investigating. Shortly afterward, he, researcher Eleanor Lacombe, and the author, with support from an internal grant from the Maine Medical Center Research Institute, founded the Maine Lyme Disease Research Laboratory.

Offshore islands are microcosms in which isolation and time establish distinct ecologies, where human independence is unfettered, and where isolated and competing animal species sort themselves out to a fragile minimum. Monhegan Island, 12 miles off Maine's midcoast, is one such. It is famous for massive headlands and an old-growth spruce forest, and is a destination for bird enthusiasts and generations of artists. It is also the year-round home of fishermen who maintain a way of life that has been the essence of the island for five centuries.

Monhegan's "Back Shore"

In 1988, the island supported about 70 year-round and 400 seasonal residents. About a third of the year-rounders hauled lobster traps. The others were sustained by tourism or were artists, carpenters, gardeners, or handymen. With limited incomes and barely enough children to maintain the one-room school, preservation of the community was – and remains – tenuous.

But because of its isolation, lack of mice and, by 1988, a crowded deer population, Monhegan offered a truly unique natural laboratory to learn more about the ecological requirements for deer ticks to become established, then infected. With that knowledge, we could explore various approaches for controlling deer ticks which might have broad application.

Martha Yandle's cat Tanya started the whole thing off. She and her husband Richard Farrell were the innkeepers of Tribler Cottage, which provided housekeeping apartments perfectly suited for four or five field scientists to spend a night or two.

Martha Yandle and Tanya

In the mid-1980s, Martha began finding ticks on her cats. At one point, entomologist Mark Wilson, one of Andy Spielman's trainees, stopped by the island on a summer cruise, apparently searching for ticks. Martha hadn't saved any, but he advised her to save any she found in a jar of alcohol and send them to the Harvard School of Public Health. Soon after, she had a collection from her cats which she dutifully sent along. Dr. Spielman's lab identified them as *Ixodes dammini.*

In 1987, the Second International Congress on Lyme Disease took place in New York City. During a dinnertime conversation among some early leaders in Lyme disease research, the subject of Martha's ticks came up. Andy Spielman was there, as were Mark Wilson and Rob Smith. From his childhood summers in midcoast Maine, Rob was familiar with Monhegan. For the diners, the questions were, now that deer ticks had appeared on a mouse-free island: Where were they obtaining their blood meals, and were they infected with the Lyme pathogen?

The upshot of the meeting was that Mark Wilson, because he was about to leave for Africa to study Rift Valley fever, would not be able to pursue the questions. That left Rob free to "have at it."

He did, making a scouting trip to the island that November which proved unpromising. Despite extensive flagging of vegetation with a large flannel cloth, he found no ticks – but then found himself marooned for two days while an ocean storm kept the mail boat *Laura B.* tied to her pier in Port Clyde.

"Flagging" is the traditional method of estimating tick abundance. The process involves dragging a square meter of fuzzy fabric, such as flannel or more durable corduroy, attached to a four-foot pole, over deer tick habitat. To gather adult ticks, which quest as high as three or four feet above the ground, we poke flags into brushy vegetation during their seasonal peaks in the fall and spring. For nymphs, which stay closer to the leaf litter, we drag the flag along the ground during their peak season in June and July.

Eleanor flagging (left). Removing ticks from flag (right).

Our first expedition left Port Clyde on the *Laura B.* on October 6th, 1988. We had invited Andy Spielman and his post-doc medical entomologist Sam Telford to join us. Pete's wife Alice came along as well, both for the ride and, incidentally, to provide delicious meals. Once disembarked, it was a short lug to Tribler Cottage and greetings from Richard and Martha. Tribler would become our island headquarters over the next two decades.

A major priority for this mission was to sample as much of the island as possible with small mammal live traps to verify that there were no mice on the island, but also to see if any other small animals (such as voles, chipmunks, or red squirrels) were present that might serve as alternative hosts for deer tick larvae and nymphs.

For the effort, we had with us around 40 American-made Sherman live traps while our Harvard colleagues came with 80 more complex but perhaps more pedigreed Longworth traps from England. These were baited with oats and peanut butter and set in places we thought looked mousy – next to suspicious holes, old stone walls, tempting mushrooms, old apple trees, even tiny fairy houses in the woods erected with twigs and moss by trusting four-year old children. After three nights not a single animal was caught, living or legendary.

With a special collector's permit from the Maine Department of Inland Fisheries and Wildlife (or IF&W), we took two deer. Their hides were quickly removed and thrust into plastic bags to avert tick escape before a more thorough examination could be conducted by Sam Telford in Boston. Disposition of the venison for island consumption was immediate. Other mammals examined during this trip included one rabbit and two muskrats. No deer ticks were found on either.

At this point, we were very discouraged by the apparent lack of ticks on Monhegan. Future trips couldn't be justified. But then, just as we were packing up to leave, Martha came down the stairs from her apartment above to say good bye. She asked if by any chance we might be interested in this bottle of ticks –some fifty of them – that she had removed from her cats since Mark Wilson's visit a couple of years before!

Baiting Sherman live traps at Tribler Cottage. Background: Dr. Spielman napping.

A week later, Sam's report on the deer hides came in. His technique was to hang the hides over water – a child's play pool works well – then fish out ticks that dropped off within four days. His news was that he had recovered 26 adult deer ticks from the two Monhegan hides. None, however, was infected with *B. burgdorferi.* This did not surprise Sam, since he had shown previously that deer are unable to pass the bacteria to uninfected ticks feeding on them. Later, we demonstrated that white-tailed deer actually disinfect infected adult ticks that feed on them.

With the usual small mammal deer tick hosts ruled out (we continued deploying Sherman traps for the next few years, completing 3000 trap nights with not a single capture), we figured that some other animal must have been providing nourishment for sub-adult deer ticks. Or maybe spring-migrating songbirds were picking them up in Connecticut, infecting them in flight, then dropping them off, for instance in the barberry bushes on Lighthouse Hill. Because a lot of the birds would have cleared Connecticut before over-wintered tick nymphs were questing down there, Andy was determined to favor the idea that a post-season invasion was occurring as songbirds on their way south picked up and infected newly-hatched larvae from points in the north.

Monhegan attracts flocks of coastal migrating songbirds in the springtime and platoons of binocular-toting birdwatchers. Martha mentioned that a woman named Margery Plymire, who had a summer cottage in close-by Port Clyde, had been banding birds on Monhegan in past years and was interested in collecting the ticks she found on them. When called, Margery eagerly accepted an invitation to join our next expedition in May of 1989. During that month, the island is alive with flocks of warblers, grosbeaks, and other canopy birds but also with ground-associating yellowthroats, song sparrows, and thrushes, species which we have shown to be reservoirs of the Lyme spirochete.

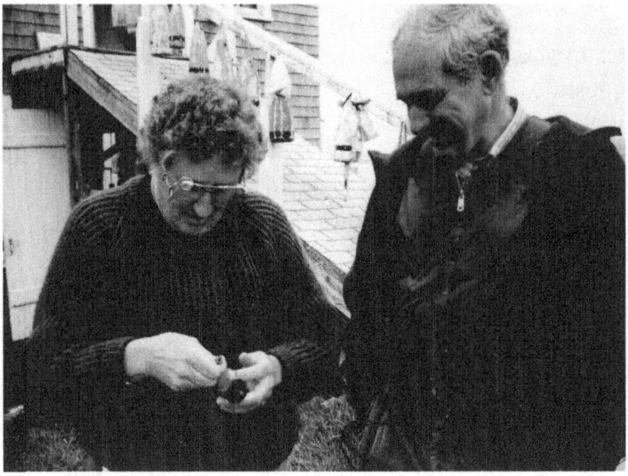

Margery Plymire and Andy Spielman

With Margery's mist nets and our boxes of live traps and baggage, getting our gear up to Tribler cottage took half an hour. We formulated a plan of action during a luncheon of Italian sandwiches, which we purchased on the ride up and extracted from a large paper bag by now weakened by leaked olive oil. In the afternoon, while others were occupied in baiting and deploying live traps around the island, Margery visited her friend Tom Martin.

Tom Martin

Tom was a jovial, offbeat photographer known as the "Birdman of Monhegan" who lived seasonally in a loft apartment in a waterfront fish house. Drawn by all things birds like to eat, his front yard was an aviary. Orioles picked away at oranges, and indigo buntings flitted about in a large stack of wire lobster traps. Margery was soon busy examining birds extracted from her nets, weighing them, gently blowing the feathers apart with a straw to assess their nutrition and look for ticks, placing a tiny aluminum tag on one leg, then filling out a detailed data sheet. Before being released, some of the birds had their pictures taken by Tom, squinting through the lens of his long-bellowed camera mounted on a sturdy tripod.

A high priority for this trip was to gather as many islanders as possible so we could discuss Lyme disease, deer ticks, and what supported their life cycle. We wanted to explain what we were hoping to do on their island and why Monhegan was unique. Helping our cause were Faryl Henderson and Luke Church, who tacked up a notice in the post office the week before we arrived. These two, along with five other EMTs and four volunteers, comprised the Monhegan Emergency Rescue Service (MERS), the island's very capable first responders. The meeting, which began at 7:30 in the school house, was well attended and had to be extended for over an hour by some penetrating questions from the audience.

As for the rest of our activities, none of our live traps was sprung. Margery found one tick (not a deer tick) on a white-throated sparrow out of 82 birds she checked, and four nymphs were found on a deer.

On our return, with calm, sunny weather, the captain favored his passengers with a side trip under the island's magnificent ocean-facing cliffs and later paused to view some frolicking phalaropes off Allen Island. But as we were debarking the *Laura B.*, an islander remarked that while there may be no mice on Monhegan, there surely were a lot of rats, particularly around the waterfront.

As the season progressed, the islanders became more aware of deer ticks and concerned about Lyme disease. With Monhegan's economy crucially dependent on income from seasonal renters and day-trippers during the ten weeks of summer, concern about contracting Lyme disease could be devastating. As a baseline measure, we felt it was important to see how many of the year-round residents had already been exposed. We were elated when they accepted our offer to be tested for antibodies to the spirochete at no cost.

Lyme flyer distributed on the island

Now supported by an internal grant from the Maine Medical Center, we returned to Monhegan in October 1989, joined by two bright young phlebotomists, Cathy Dragoni and Judy Wicken. Their gear – syringes, vials, a centrifuge, forms, coolers – was set up by the meat counter in the Monhegan Store, with the gracious consent of its proprietor, Jack Bell.

Cathy Dragoni drawing blood in the Monhegan Store

At three o'clock the residents began to gather, and within 15 minutes the line extended from the cash register to the back of the store. Kids came screaming downhill from the school house on their bikes to see what was going on. Cathy and Judy drew blood for two-and-a-half hours from 55 residents, which was an exceptional turnout for a community of around 70 souls. There were very few questions and a lot of enthusiasm and humor.

At the end of the afternoon, the phone rang. Amid chaos, Jack answered it. He reported that the Lyme disease crew were all invited to the Fish House for a beer. After hauling the gear back to Tribler, we went to the Fish House where a bar and festive appetizers were set up, but only one or two people were present. We thanked them, stayed a little while as others joined the group, then went back to Tribler for our own refreshments and supper. Later, it turned out that to our embarrassment, Jack had misinterpreted a request for the Lyme disease people to go to Sherm Stanley's house to pick up a deer they had shot for us as an invitation to go to the party for a beer. This left the people at Sherm's disgusted at our lack of interest in their hunting efforts and the people at the Fish House wondering who these off-island party crashers were!

Only one of the blood samples turned out to be positive for antibodies to the Lyme bacterium, that from a young girl who, just after our visit, developed a hot, swollen elbow and a fever and felt miserable. Rob made the diagnosis over the phone and arranged for her antibiotic to be delivered to Port Clyde overnight. As is often a problem on islands unconnected to the mainland with a bridge, however, delivery to the patient was delayed several anxious days while the *Laura B.* waited out an ocean storm. This was Monhegan's first physician-diagnosed case of Lyme disease.

We flagged 34 ticks on this visit, and when the hide on the Stanley's deer was discretely hung for four days in the author's barn in Cape Elizabeth, another 18 adults fell into a tub of water below. All but one of these were males, a typical ratio since during the adult deer tick season, males crowd onto deer not to feed but to mate repeatedly with females. Having succeeded, they remain to prowl for further opportunities, but the mated females drop off as soon as they have had their fill of deer blood, hence the gender gap.

When Sam examined our flagged ticks in his lab later, he found the Lyme spirochete in 23 percent of them. We now had solid evidence that deer ticks were well-established on the island, were infected, and were infecting people. We were eager for more ticks, so to increase our take we left a poster offering a dollar for every live tick we received from the island.

We also wanted to find out how effectively the ticks were transmitting the Lyme spirochete on the island. For that, we offered the owners of all dogs and cats a free blood test for their pets to see if they had been infected by a previous tick bite. The offer was widely accepted and a date in the middle of May was set. But space was limited.

A boisterous pop-up animal clinic did not seem appropriately hygienic for the Monhegan store nor for the school house, but a venue was finally found in a space the islanders used for exercise work-outs. This was the cramped second floor of the Island Spa, an older building which at the time housed a tourist shop on the first floor. For blood drawing, we enticed the state veterinarian, Dr. Chip Ridkey, to join us, and he brought along a colleague, Dr. Don Hoenig. They had a rough trip out; it took a while to get their color back to normal.

These two professionals were clones of Hawkeye Pierce and B. J. Hunnicutt from the TV show "MASH." In three bleeding sessions from 1 pm through 7 pm, the owners of 30 cats and 13 dogs clambered up and down a steep, narrow back staircase with their charges, many of whom were socially challenged. While owners registered and jotted their way through a short questionnaire, fur flew in the background with yelps, snarls, and howls, and the two veterinarians bantered hilariously. The high point came when a bent, bearded lobsterman emerged from the staircase lugging a good-sized crate which bobbed heavily and emitted violent scratching noises. "So who do we have here?" asked Eleanor behind the table. "DEATH ANGEL!" came the answer. It took both vets to subdue the 20-pound feline. They couldn't get a word out between gritted teeth.

The results from our matinee were impressive: one-quarter of both cats and dogs were positive for Lyme antibodies.

Although the pet survey was the primary mission for this trip, we brought along 11 larger Tomahawk live traps to see if we could capture some of the rats the islanders had mentioned at our last visit. At the time, medical entomologists did not know if Norway rats served as hosts for deer ticks, if they could become infected with Lyme bacterium, and if so, could they could pass it on to uninfected ticks feeding on them; that is, were they an important natural reservoir? During a break in the pandemonium at the Island Spa, the traps were hurriedly set around the fish houses, the generator house, and next to the dumpster on the pier. By the next morning, one of the traps had mysteriously disappeared, but none of the remaining ten had been disturbed. This was an inauspicious start to what would become the major thrust of our activities on the island over the next three years.

In the meantime, having found Lyme-infected deer ticks on an island lacking their recognized reservoir hosts, we had an enigma to share with the world of medical entomologists and tick ecologists. Rob was invited to present a poster entitled "Presence of *Borrelia burgdorferi* infection in *Ixodes dammini* Ticks in the Absence of Mice" at the Third International Conference on Lyme Disease in Stockholm, Sweden in the fall of 1990. (It was at the previous International Conference in New York that the Monhegan project was launched by Rob's chat with Mark Wilson.) Among attendees were scientists from the U.S. Center for Disease Control's Vector-borne Disease Laboratory in Fort Collins, Colorado. They recognized that the basic question we presented needed resolution. A few months later, the CDC requested research proposals on this aspect of tick-borne diseases, the requirements for which could only be accomplished on an island like Monhegan. The forthcoming grant funded much of our work on the island in the early 1990s.

Norway rats were the prime reservoir suspects. Given the island's history of European exploration and trade dating back to the 1600s, these rodents had undoubtedly infested the island for centuries. Our first task was to live-trap as many as possible, to examine each one for deer tick larvae and nymphs (adult ticks don't get on rats), and to take a tiny piece of skin from an ear on each to see if we could grow the Lyme spirochete from it back in the lab.

Norway rat in a Tomahawk Trap

For this we needed a trained field biologist on the island during the summer, a job happily accepted by Steve Rich, a post-graduate biology student from the University of Vermont. Steve trapped for us on Monhegan for three summers, integrating himself into the island community (when we called the phone number he gave us, it was answered by a very pleasant young lady). His first summer's efforts yielded 31 Norway rats that were heavily infested with deer tick larvae and nymphs. Lyme bacteria were found in over 60 percent of both stages. In addition, in the lab we were able to culture *Borrelia burgdorferi* from the rat ear samples.

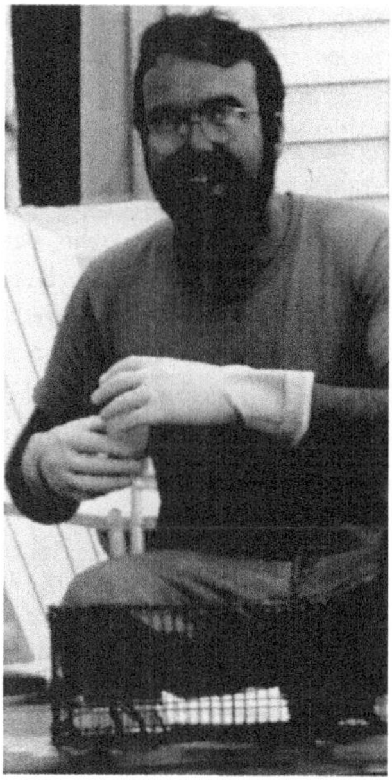

Steve Rich, the "pied pier" of Monhegan

Newly hatched deer tick larvae are uninfected: the spirochete-laden larvae removed from rats gave ample evidence that this species was very competent in transmitting the pathogen. This was news. Norway rats had never been recognized as competent reservoir hosts for the Lyme pathogen. Another poster was presented, followed by a paper.

Throughout our experience on Monhegan, through meetings, letters to the community, and via numerous articles in *The New Monhegan Press,* we made sure that the islanders knew of our activities, findings, plans, and interpretations. Peter Boehmer and his wife Raquel published this "nearly monthly" newsletter which chronicled the citizens' business, featured expressive contributions from its readers, and pulled few punches reporting on island controversies and attitudes. As a result, particularly from our first meeting with the islanders in the schoolhouse, the critical role of white-tail deer in the tick's life cycle began to be appreciated, particularly by the Monhegan Associates who own and protect two-thirds of the island.

This organization was founded in 1954 by the son of Thomas Alva Edison with a mission to preserve the island's "natural beauty, biotic community, desirable features, and the simple, friendly way of life that has existed on the Island as a whole." Members were particularly concerned about the damage the burgeoning deer herd was causing to the island's ecology. Their severe browsing had caused deer-resistant exotic Japanese barberry to proliferate; deer also decimated islanders' gardens and plantings. But Lyme disease was the straw the broke the camel's back. From the *New Monhegan Press*, Vol. II, No. 6, September 1990:

> **Monhegan Associates Meet**
>
> The deer removal debate was a repeat of the night before. Even so, the Ecology Committee insisted that the Trustees vote on the matter, and that they did, with 3 to 4 in favor of removal of the deer in conjunction with the Plantation (i.e., the town would have to vote on the matter). Five Trustees abstained from the vote. As someone said prior to the vote, a positive vote would make no difference as the Town would then vote it down.
>
> ED. Note...*Perhaps. But such an action will be very much more serious now. The reason is that the Associates have gone on record in attempting to reduce a threat to ecology and to the risk of health to the community. (one might debate the level of the risks, but one may no longer question their present reality). Should the Town elect to keep the deer, they are, in effect accepting the responsibility for any future manifestations of such threats. Clearly, the ball (deer) is now in the town's court.*
>
> *While the relative threat the deer represent is being debated, what the relative loss of the removal of the deer represents, has not been articulated. Another way of asking the question is: We know there are risks to keeping the deer – what are the risks (or losses) of having them removed, and who are the losers?*

Despite the Associates' broad membership comprising both year-round and seasonal residents, the Plantation, at its next Annual Meeting, voted not to go along, as reported in *The Island News* of May, 1991:

> **Monhegan Votes to Keep Its Deer Herd**
>
> The removal of island deer was a controversial issue at this year's [annual town] meeting. Acting on a recommendation introduced from their ecology committee, the Monhegan Associates who own and maintain Monhegan's "wildlands" introduced an article to petition the state to remove the deer herd from Monhegan. The committee cited the loss of about 59 species of island wildflowers since the introduction of deer by the state in 1954. Concern was also expressed about the increasing danger of Lyme tick disease to the community. The article was rejected by a vote of 21 to 10.

The components of what is called an integrated approach to control deer ticks include clearing away the vegetation they thrive in and spraying with tick-killing acaricides. As a result of decades of browsing native understory however, what remained throughout Monhegan's square mile were lush, impenetrable thickets of well-established Japanese barberry, prime habitat for deer ticks but impossible to eradicate. It would be equally foolhardy to apply pesticides, as the town's primary water supply comes from the "Meadow," a nine-acre peat bog abutting the village. Besides, vehement resistance from lobstermen against any compound that might impact their livelihood would be guaranteed.

But the risk to the islanders' health was increasing. The number of ticks we could flag had doubled, and now four in ten were infected. With removing deer off-limits, the only remaining approach to control ticks and reduce disease transmission involved the rats. In 1992, we devised a plan that would first determine where the rats were nesting on the island, then see if we could lower the tick population by attracting them to bait tubes in which they would be dusted with a tick-killing compound. If that approach failed, we might attempt finally to lower their numbers drastically with a concerted rat control program. In July, both the Plantation and the Associates voted to have us proceed.

The rat distribution project was carried out diligently by the young namesake nephew of one of the island's stalwart old-timers, Doug Odum. First we overlaid the island map with a grid that divided it into 140 quadrants, then constructed as many bait tubes. Each was two feet long, sawn from eight-foot long, three-inch diameter PVC pipe in a Cape Elizabeth basement. At the middle of each, a one-inch hole was provided to admit a cork. Next, a rack was made to hold 96 plastic 35 mm film canisters open-end up. Each of these received a squirt of PAM.

Bait tube shop (left). Bait with nails solidified in film canisters (right).

For rat bait, we set three blocks of paraffin in a steaming double boiler on a domestic range. In another double boiler went two cups each of peanut butter, shredded suet, Quaker oats, raisins – plus, since these were working waterfront rats, a can of sardines. After we upended the molten paraffin into the bait mix and stirred thoroughly, we poured the contents, carefully, into the 96 canisters. Next, we placed a ten-penny nail point up in each canister so that, after solidification and removal, the bait could be attached to the underside of a cork. Just before use, we placed the canisters briefly in hot water, punctured a hole in the bottom of each to relieve suction, and slipped the molded bait cylinder out.

Unloading bait tubes in Port Clyde

It took Doug several days in late October 1992 to distribute the tubes around the island. Then he spent November repeatedly returning to each one, removing the cork and recording how many times he found the bait cylinder untouched, nibbled, partially eaten, or totally consumed. When his pages of data were summarized on the grid map, they revealed that rats were present in patches over most of the island, with heavy infestations beside the Meadow and above Lobster Cove. This was the information we needed to plan our attack.

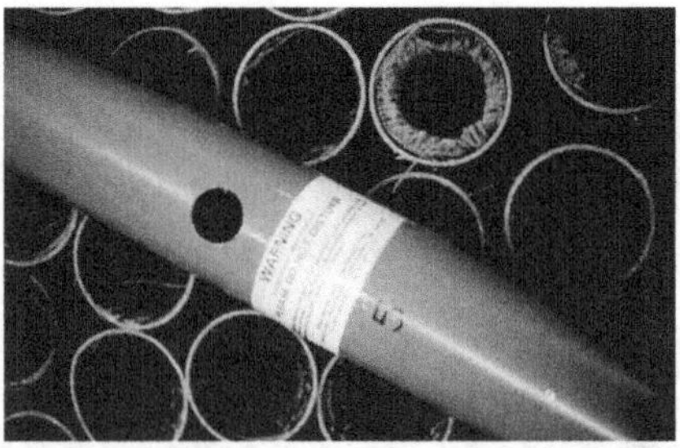

Rat bait tubes with rug-lined entrance and bait/cork hole

But the attack lacked a crucial weapon. One of the contributions to reducing the risk of Lyme disease from Andy Spielman's lab was a simple but brilliant device to trick white-footed mice to coat themselves with a tick-killing powder in their nests. It consisted of a short cardboard tube stuffed with pledgets of cotton batting that had been treated with permethrin, an efficient tick-killing pesticide. These would be sought by mice to line their nests, where the compound would kill ticks for as many generations as were swaddled by the cotton. Andy named the tube Damminix, which remains one of the tools for the integrated management of deer ticks. We had very effective rat bait tubes. Why couldn't we use them to coat rats with permethrin?

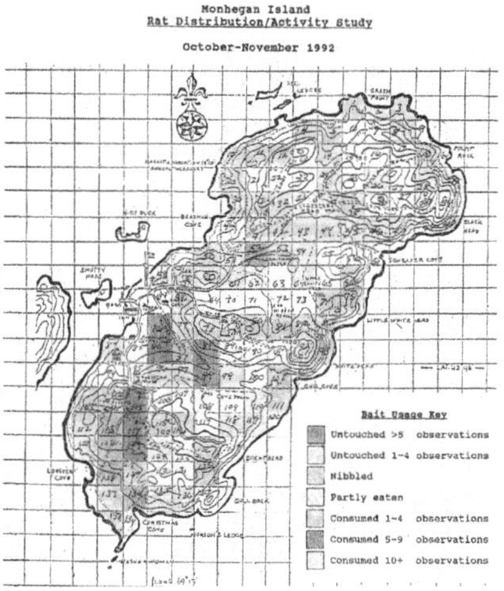

Grid map of rat activity on Monhegan

So a number were modified by gluing strips of shag rug around the inside of each end. Once on the island, we coated these with permethrin powder and set them out for several nights within one of the high rat areas. Then we trapped rats both within and well away from the treatment area and examined them for ticks. To make sure the rats had been effectively treated, we added a yellow-fluorescing marker to the permethrin which would be visible when the dusted rats were examined in the dark with an ultraviolet light.

We were elated when, in our lodging's cellar, our blue UV light induced yellow fluorescence in the fur of a tick-free rat from the treatment-area. But our celebration was short-lived. The project hit an unforeseen snag a couple of days later when each bait tube became overrun with slime-exuding slugs, rendering them worse than useless. It was equally frustrating when we began to see yellow fluorescence on rats from the remote area, only to find subsequently that rat urine fluoresces yellow!

The season was over. Because it was important not to distract the rats from alternatively-purposed bait stations in the spring, the project died.

Rat control and rat eradication projects have been attempted on many islands around the world, primarily to prevent the extinction of nesting seabirds, turtles, and other endangered fauna. The usual attempts have involved distributing toxic bait in rat habitat. Where eradication has been successful, however, the method used – repeated dropping of poisoned bait from aircraft – would be inconceivable over human habitation and dangerous anywhere to non-target animals.

Nevertheless, our goal was to eliminate the rat population on Monhegan, to cut off the essential blood meals for sub-adult deer ticks, and to get rid of the reservoir of spirochetes threatening all the humans and canines on the island. This was an incredibly long shot, but given the villagers' preference to preserve the deer, it was the only island-wide tick control approach that might work.

Preparation for the project occupied all of 1992. To proceed, we needed the Plantation and the Associates on our side as well as permission from three state agencies (The Department of Inland Fisheries and Wildlife, the Board of Pesticide Control, and the Land Use Regulation Commission), and most importantly, from the U.S. Department of Agriculture's Animal Damage Control Branch. ADC had to be the boots on the ground to deploy a rodenticide called zinc phosphide in open, non-residential areas. The other part of the plan was to provide an ostensibly safer compound, Quintox, to villagers who would be instructed on its proper application.

Rats Are Having a Hard Spring

If winter weren't hard enough, spring is more difficult for the Monhegan rat population. We are told that the rats are just about out of their winter food supply and on the point of starvation. The females need high protein for the breeding cycle. Not only was March a terror as far as their food needs go, the first half of April was no better. Perhaps that is what accounted for the low rat activity some of us encountered in working with experts from the State and Federal agencies to reduce the rat population on the island.

The New Monhegan Press, April, 1993

The plan was to make our assault at the end of winter when the rats would be particularly hungry and before too much procreation had taken place. But plans to go to the island in late March were thwarted twice by late winter snowstorms.

Ed Butler, Maine's Animal Damage Control Officer, took charge of the project. Along with Ed, a colleague from New Hampshire's ADC, Marsha Barden, and Ed's boss, Dennis Slate, we landed on Monhegan on April 5th, 1993.

The weather was beautiful, but much of the island was covered with heavy drifts of slushy snow. That was not a good sign. The material explaining the project that we had previously provided to the islanders contained a solicitation for homeowners who were willing to place Quintox in concealed places around their buildings. Eight stalwart volunteers came to a meeting we held in the schoolhouse that evening. After we explained the program and how to avoid risks to their pets, Ed showed them a video provided by the USDA which was mostly irrelevant to Monhegan's situations. Before leaving, the Quintox was distributed with yet more instructions along with record sheets that we banked on their filling out.

The next morning Ed and Marsha mixed the zinc phosphide with peanut butter and oats, and then with Doug Odom in the lead they set off to place baited tubes around the island. Their work was frustrated by heaps of melting snow.

Ed Butler mixing rat poison

More outreach and education followed later in the day as Ed lead the school teacher and her charges around the village showing them where rats live. There was no doubt about his fervor when he and the Boehmers' Jack Russell terrier Twine competed to see who could unearth a rat under a boulder first.

Rivals for a rat

Twine, Bailer, Diesel, and Badger were four of the feisty terriers on the island at the time whose presence could be traced to a respected frequent visitor and champion of the breed, artist Jamie Wyeth.

Homer, by Jamie Wyeth. With kind permission of the artist.

Unfortunately, the homeowner-dispersed Quintox project was brought to a hasty halt by the innate predator traits of the Jack Russell terrier. While the bait could be placed where dogs could not get at it, the same did not apply to rats that consumed it. Indeed, Twine was witnessed dispatching a rat in the vicinity of some hidden bait, and shortly thereafter began to display alarming symptoms requiring an urgent trip to a mainland veterinarian. His eventual full recovery was not immediate, but recall of all Quintox was.

Shortly after, we heard through the grapevine that, at a meeting of the Monhegan Associates at which our project was discussed, a senior officer was heard to say "What are these people doing, anyway?"

Nor were the poison bait tubes deployed by our Federal colleagues effective. When Steve Rich set live-traps nearby the bait stations two months later, he captured 15 rats. From nine of these he removed 88 deer ticks, all but one infected with Lyme spirochetes. Although our USDA collaborators and many of the islanders hoped to see the program continued indefinitely, funds could not be raised. The effort ended with islanders being provided free access to a large supply of rat snap traps.

It was pretty clear to everyone that rat control, as a method to reduce Lyme-carrying ticks, was not going to work. It was time to review what options remained. We composed a letter summarizing the attempts we had made to control ticks on the island that did not involve deer. We ended it by describing what appeared to be an effective, novel approach. We sent it to each household on the island and to Peter Boehmer to inform the much larger community of seasonal residents who subscribed to his publication.

The idea of reducing the deer herd was still a very sensitive issue, but we were now able to suggest an approach that had proved effective in controlling lone-star ticks infesting white-tailed deer in Texas. It involved baiting deer to feeders filled with grain treated with ivermectin. The drug is used world-wide to kill a variety of human parasites and to prevent heartworm infection in dogs. Given a sufficient concentration in a deer's blood, a deer tick feeding on it will die. Rather than culling deer, we could use the animals to kill the mated female ticks that were feeding on them. The idea was quickly accepted in the village.

But it took most of 1994 the make the arrangements. Thankfully, we gained the interest of the USDA researchers in Texas who had developed the approach. What may have been persuasive as well were an isolated deer herd, a major disease vector tick not found in Texas, increasing cases of Lyme disease in an island community, and some experienced tick ecologists to do the ground work. The Texans' collaboration was critical because they had already obtained the experimental drug use permits from the USDA and FDA, so they could supply us with the drug, provide laboratory services to determine ivermectin levels, and share invaluable advice on doses, distribution, and feeding strategies. Our job was to win the approval of Maine's Department of Inland Fisheries and Wildlife, the Land Use Regulation Commission (as an unorganized township, the Monhegan Plantation was administered by LURC), the Bureau of Pesticide Control, and, critically, the residents of Monhegan.

In January 1994, we crafted an article for consideration at the annual town meeting in March :

> To allow Maine Medical Center investigators, with support from the USDA, to undertake a program to reduce the risk of Lyme disease on Monhegan by treating deer with an environmentally safe tick-killing compound, pending receipt of all necessary permits.

It passed, but not by much. The main problem was a stipulation in the FDA permit requiring that no venison be consumed for 90 days from the last day a harvested deer might have eaten any of the ivermectin-laced feed. That, in turn, meant that IF&W had to close Monhegan to deer hunting, but that could not be accomplished until a formal request from the Plantation was considered and passed by an IF&W Policy Board which met in August. Passage of the article in the town warrant would have allowed Katie Boegel, the Chief Assessor (select person), to send such a letter. The Assessors, however, wished the whole deer issue to be reviewed at a public meeting, which was hastily arranged so that if a letter were approved, it could be sent to the State by August 1 in order to initiate a process that required 45 days.

On July 27 we met on the dock in Port Clyde with three wildlife biologists from IF&W: Gerry Lavigne, the state's deer biologist, Gene Dumont, biologist for the

Monhegan region, and Ken Elowe, who was in charge of IF&W's mammalian biologists.

Gerry Lavigne. Photo used with permission. Photo credit: Jeannine O. Lavigne

Gerry was a straightforward, highly principled wildlife biologist, now in his twentieth year at IF&W, almost all of them passionately devoted to deer management. He was the leader of one expansive project modeling deer populations on the basis of biological data from over 30,000 harvested deer over a five-year period, and a second project that predicted annual deer populations based on an index of winter severity. Those predictions had become the mainstay of the Department's deer management strategy. IF&W is keeper of Maine's deer, and nothing relating to deer can be done without its approval, which frequently involves review at multiple administrative levels. Gerry shepherded us through the permitting process and became our liaison with his agency and an authority on deer at multiple meetings with the citizens of Monhegan Island. Without him, our project could not have been completed, if started at all.

The trip out to Monhegan was warm and hazy. After lunch there was time for the biologists to appreciate the habitat and topography of the island. Later, 30 to 35 islanders crowded into the schoolhouse for the deer meeting. Some strongly resisted deer elimination; others strongly desired it and were critical of the State for not assuming responsibility. There was much worry that our deer feeding would lead to an increased deer population, and several complained about the 90-day consumption ban. Participation by the IF&W crew was extremely useful. Ken Elowe's explanation of the biological aspects and statutory requirements were particularly to the point. Unfortunately, there were too few Assessors present to vote on sending the request letter to IF&W

then and there, but a meeting date was set for the following Saturday to conclude the matter.

The meeting was held and the letter was sent in time. While we waited for it to be considered, we had to address a remaining truly critical hurdle: how to pay for the project. An earlier application submitted by the Plantation to the Maine Community Foundation appeared to have staff support initially but was subsequently denied without comment. Fortunately, a second, submitted to the Bingham Betterment Foundation, won approval, and it supported the project for the next two years.

The plan called for establishing feeding stations around the island which would be kept filled with special twice-washed corn pre-mixed with a specific amount of ivermectin liquid. Over a two-year period, there would be 16 stations. Eight were troughs made voluntarily on the island by Charlie MacDonald; the remainder were a growing number of automatic feeders supplied by our USDA collaborators. They would spew about seven pounds of treated corn 30 feet, at first at dusk; later, twice a day. While the troughs were placed near the homes of motivated residents, the automatic feeders were set up at remote sites requiring sherpas, mainly Luke Church, to lug in 60-pound knapsacks sometimes twice a week. Feeding was continuous during both the fall and spring seasons of adult deer ticks from late September of 1994 through November of 1996.

Adding ivermectin to corn (left). Corn dispenser (center). Luke Church (right).

There were three ways to measure the effectiveness of ivermectin delivery as we went along. First, half way through each treatment period we picked up deer pellets all over the island and sent them to our USDA colleagues. They had a bioassay which involved measuring the ability of the scat to prevent the development of hornfly larvae (ivermectin remains viable after being eaten and digested.) The second was to measure the compound directly in the blood of deer taken during the feeding periods, and the third was to count the number of ticks that fell off hides hung above water in the author's cellar, measure their weight and size, and measure the success of the fed females to lay eggs and those eggs to hatch.

Author collecting pellets freshly provided by Honey, a village pet.

Deer blood analysis showed that when the deer ate enough corn to get their blood ivermectin levels above 15 nanograms per milliliter there would be very few – and smaller – female ticks feeding on them and fewer still that deposited eggs and eggs that hatched. In addition, there were significantly fewer ticks on the hides of Monhegan deer with ivermectin levels above this threshold in comparison with ticks on deer from Isle au Haut, our "control" island (more about that later).

The problem was this: only a little over half the deer reached the 15 nanogram threshold, so that by the end of two years, we could not see a difference in the number of ticks we could flag in the fall. There were a couple of reasons: first, we had underestimated the number of deer on the island and therefore, despite the 14 tons of corn lugged to the feeders, it wasn't enough; and second, the larger bucks hogged the feeders, preventing more timid deer from consuming the medicated corn. Also, there was thievery. Faryl Henderson discovered a rat warren that overflowed with corn stored for the winter.

We encouraged the islanders to consider one more year of treatment if we could find funding, but we had run out of time. They (though not all of them) wanted to reduce the island's deer herd. Nevertheless, we learned a lot from the experiment.

For those wanting to preserve the herd, there was one more option: immuno-conception. It had been successfully used by the Humane Society of the United States (HSUS) to control a population of wild horses in a Maryland preserve and on fence-enclosed white-tailed deer elsewhere. The method requires injecting does with a vaccine that causes their immune system to destroy a coating on ovarian eggs, preventing them from becoming fertilized. If most of the does could be treated (it took two vaccinations), the deer population might stabilize, but it would be a long time before a reduction sufficient to control tick numbers could be expected. Despite the daunting challenge of accomplishing this in Monhegan's wildlands, HSUS principals Allen Rutberg and Rick Naugle expressed interest in making the attempt when they visited the island in May, 1996. After a trip around the island they agreed that the project could not be carried out on the rugged, entangled, northern half of the island, but perhaps it might work in the more open southerly half. The matter was settled, however, when an application was made to IF&W. Their response:

> The HSUS/MMC proposal was reviewed by Wildlife Division biologists who recommended denial of the scientific collection permit amendment which would enable this proposal. Specifically, Wildlife Division staff had many concerns regarding study design as well as social implications of intensively handling wild deer to the degree necessary to control herd growth. Immunocontraception, even if successful, would not quickly reduce deer numbers or the immediate Lyme disease threat. In short, this approach offered no long-term solution to the problem and had little chance of success.

Thoroughness demands mention of one further possibility: using guinea hens to control deer ticks. It had been widely (albeit anecdotally) reported that in the process of pecking at everything in their path they preyed on ticks. For a short time in the mid-1990s, there was a free-range flock of eight of the incessantly calling hens around the village, so someone asked if perhaps more were needed on Monhegan. The birds were not popular. On one occasion while we were on the island, a guinea hen that died of an undisclosed cause was presented to our new field biologist, Chuck Lubelczyk. He examined it carefully, then performed an autopsy. There were no ticks in the bird's gut, but nymphs were found under its feathers. That particular guinea hen was feeding, not killing, deer ticks. Fortunately, hypothetical tick control did not save the fowl's presence on the island.

Chuck followed Steve Rich as our field biologist. He came to us highly recommended in 1995 fresh out of a wildlife biology program at the University of New Hampshire where camping out in the winter in order to track deer was part of his portfolio. That experience came in handy on Monhegan where he had to pitch his tent during his solo rat-trapping expeditions. The Monhegan Associates didn't allow camping on their land, but artist Lynne Drexler let him use a patch beside her house on the hill behind the village. Not a bed of roses; the first time out he was flooded out in a deluge and his food supplies were soaked – all but the rat trap bait, peanut butter, and sardines, which he subsisted on until we could mail him some cash. Once dried out, he remembers waking up with the warm sun and seeing the shadows of deer ticks climbing up the outside of his tent. That was twenty years ago.

Chuck Lubelczyk weighing Monhegan rat

In the meantime, Chuck has become a skilled vector ecologist representing the lab nationally, contributing papers to the scientific literature and, by developing a network of mutually beneficial collaborations with state and federal epidemiologists, scientists, and other professionals, has expanded the scope of lab's activities and enhanced its stature well beyond the state's borders.

Finally, on December 9, 1996, the Monhegan Plantation held a special meeting to determine the fate of the island's deer herd. Afterwards, Mary Beth Dolan, First Assessor, wrote to Gerry Lavigne:

> Dear Gerry:
> On December third at a special town meeting the residents voted to ask the Department of Inland Fisheries and Wildlife to facilitate reduction of the deer herd on Monhegan Island. The actual wording of the article is as follows: "To see if the residents will vote to ask the Department of Inland Fisheries and Wildlife to facilitate the reduction of the deer herd on Monhegan, under the Maine Medical Center's scientific collector's permit to 15 or fewer deer due to the significant threat of Lyme disease." The vote was 22 yes and 7 no.

CHAPTER 2.

ISLE AU HAUT

By 1989, it had become obvious that we were committed to investigating and addressing Monhegan's unique mouseless deer tick invasion. We then needed to find a control island, similarly isolated, but with deer, ticks, *and* mice. Isle au Haut, fairly isolated yet served by a ferry from Stonington, had lots of deer and mice. But did it have ticks? This was early in their infiltration into Maine. They were reliably established along the south coast, but not beyond.

Since nearly two-thirds of the island belonged to Acadia National Park, the first priority would be to gain the interest and possibly the collaboration of Park officials. Our early meetings with them were successful on both counts, perhaps influenced by the fact that both the Park's new director and its biologist had just come from Fire Island National Seashore on Long Island, New York, where ticks and Lyme disease were rampant. Arrangements were made not only for a trip to Isle au Haut, but also for collaboration on a two-year tick survey of Mt. Desert Island conducted by their staff biologist, Judy Hazen Connery.

The plan was to go out in a Park Service boat from Southwest Harbor to the island where we would be joined by Dr. Kenneth Crowell, Professor of Biology at St. Lawrence University and an expert on small mammal populations on islands. Ken lived on Deer Isle and conducted much of his work on the islands in the area. This was his territory, and he would be an invaluable mentor and collaborator.

The run was a little over 20 miles. Altogether with Park personnel, we were seven on board. From the trip record for June 7-9, 1989:

> It was a gray morning with light winds as we moved out Western Way at about twenty knots. Abruptly, however, as we were passing over Bass Harbor Head bar, the captain shut the engines down, concerned about a loss of power and black smoke coming out the exhaust. We circled a bit until he decided that he couldn't proceed any further and turned the vessel back to Southwest Harbor at steerageway speed. Things looked quite grim because we had made arrangements with the

Crowells, who live on Deer Isle and were going to travel independently in their own boat, to meet us at Duck Harbor on Isle au Haut. Thanks to hand-held radios, a phone call was placed and Dr. Crowell was stopped just before getting on his boat at Sunset. An hour later we were back in Park headquarters making arrangements with the Crowells to meet them in Stonington where the Park Service had chartered the mail boat *Mink* for a special trip. Once on the island we met Kent Mattingly, a tall, thin, gentle and articulate National Park ranger who, with two trucks, transported us a mile or two south to Eli Creek. Five minutes down a trail was a single room camp, about 15′ x 20′ with two superimposed bunks, a mattress in a loft, a propane stove, sink, table and a few chairs. Once at camp, haste was made to set out traps since only three hours of light remained. The weather was gray and threatening showers. When everyone was back, three tents were set up, two by the shore and one above camp. During the night it rained and things became generally soggy. In the morning rain fell in earnest.

Eli Creek campsite

In all, 15 mice and one shrew came in, allowing Ken to show us the process he had developed to examine them. First, the animal is shaken out of the trap into a plastic bag, best the long skinny ones for bread loaves or the morning newspaper. After being weighed by hanging the bag from a spring scale, the researcher extracts it by the tail with one hand and lets it get a foothold on their thigh in order to secure it by the scruff of its neck with the other hand. Thus immobilized, the mouse is examined for body and tail length, sex, pregnancy, age, and ticks. If handled gently, mice are usually quite cooperative. But all this works better in the light of the day when it's not pouring rain.

Telling the difference between a white-footed mouse (*Peromyscus leucopus*) and a deer mouse (*P. maniculatus*) can be challenging. The deer mouse's tail is longer than its body and white underneath. That all these appeared to be deer mice was important to us, since this species had never been shown to host deer ticks. But this time no ticks were found on any of them, although many were infested with fleas.

Boom Beach, Isle au Haut

Shortly after this trip, Marcus Russell, who had been trained in wildlife management, became our field biologist. One of his early assignments was to explore Isle au Haut with a flag and Sherman traps to see what he could find for deer ticks. Marcus set the traps out in a 7×7 grid pattern on a town reserve known as Boom Beach, a successional field on the east side of the island sloping to a stony shoreline with bayberry, huckleberry, juniper, a few barberry bushes, and crossed by deer trails. He was concerned because despite good habitat, he flagged no ticks. But then, when he sat down to have a sandwich for lunch, he noticed the ground he was sitting on was crawling with larvae!

In addition, he found deer tick larvae on many of the 70 mice he captured on Boom Beach, and when examined back in the lab, we found several teaming with spirochetes. To top things off, a little later we learned that Lyme disease had been diagnosed in two dogs on the island. Isle au Haut not only supported deer ticks, they were infected, and they were transmitting Lyme disease.

We then had our study site, but needed to find a camp nearby for overnight shelter. Ted Ames, long-time lobsterman and fisheries conservation scientist, and his wife Robin Alden, who subsequently became Commissioner of the Department of Marine Resources, had a camp near an indentation on the east shore known as Sheep Thief Gulch. We contacted Ted, who said his camp was sometimes occupied on weekends, but saw no problem if Marcus and Steve Rich wanted to put up a tent up nearby. We left it that the two would be out the next Tuesday to stay through Friday.

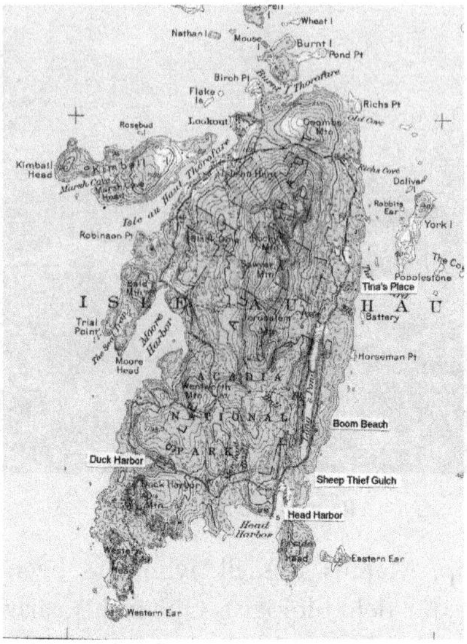

Map of Isle au Haut

Somewhere along the line, the bit about tenting didn't get through to our field staff, who took it upon themselves to innocently get into Ted's locked camp through the rear window and spend the night!

Dr. Ted Ames
July 12, 1991
Alden/Ames Laboratory
Rte. 15, Stonington ME 04681

Dear Ted:
Not an auspicious way to gain the support of landowners at a critical study site! We hope that, to some extent this package will make amends for what was an honest misunderstanding on the part of my otherwise extraordinarily competent field workers……

With ample deer, mice, and solid evidence that deer ticks were established, Isle au Haut was a perfect control island for Monhegan, but we needed to find both a place to stay and, the island being five miles long, transportation to explore it. It was a godsend that Tina Tulley, an involved islander concerned about emerging Lyme disease, offered to provide both.

When we went out in October, Tina put us in her "cottage," a two-story, three-bedroom shingled summer home with full kitchen, living room with beach-stone fireplace and formal dining room. The house was called "Ardensea."

Ardensea

Over the next decade, we had either Ardensea or a neat, fully equipped camp above the steep easterly shore. Tina also generously allowed us to use one of her cars. Offshore Isle au Haut has no garage to perform annual inspections, so very delicate driving was required, allowing lots of time to brake and to nudge over rocks and potholes. We could not have done Isle au Haut without generous, committed Tina Tulley.

Steve Rich also helped out on the island. As part of his Master's Program at The University of Vermont, he was studying the molecular genetics of voles on islands. Fortuitously, one byproduct was a method by which deer mice could be differentiated from white-footed mice by analyzing an enzyme in their saliva. Thus, the restrained mice had to put up with one more insult before being released: having their mouths flushed out with a 1 ml syringe. Steve was able to demonstrate that all the mice captured on Isle au Haut were indeed deer mice, *Peromyscus maniculatus,* evidence for a subsequent paper adding deer mice to the list of vertebrate hosts that both feed and infect the vector tick of Lyme disease.

Isle au Haut was not through contributing to the scientific literature on tick-borne diseases. In 1994, a couple who lived by the church sent us in a tick for identification they took from their one-month old kid goat.

> Discovered tick because kid was making short distress calls. Tick very engorged, ear very sore to touch. By 2 pm kid's ear, neck, and shoulders all stiff and very sore. Frequent crying, standing looking miserable and trying to lie down for short periods. 6 am swelling worse, ear flopping down with weight of swelling but acute tenderness gone. Temp 103°, acting, happy, eating well.

The tick was a mouse tick, *Ixodes muris,* now becoming rare. This was the first of several cases of domestic animals that also developed high fevers and massive swelling after bites from this tick, a previously unrecognized veterinary condition the cause of which remains unknown.

One unofficial visit to the island was in the author's sailboat, our plan being to anchor in Head Harbor at the south end of the island, then hike the short distance to check out the Boom Beach study site. Our approach turned out to be buffeted by rain and heavy headwinds, but we were relieved to find a quiet spot in the harbor to anchor, after skirting around a herring net set in the middle.

Dr. Rob Smith off Western Ear, Isle au Haut

Quiet for a while. Just as we were sitting back enjoying the sunset with hints of supper wafting up from the galley, a fusillade of gunfire scrambled us all down into the cabin. There was a barge moored near the net, with some tarps strung up near the stern to make a tent. Squinting through the binoculars with our heads down we could make out, from time to time, a form emerge with a rifle and start blazing away at gulls and cormorants on the water, some of the rounds ricocheting off the water too close for comfort. It was getting too dark to see. We stayed down below. No shots hit us. Eventually darkness returned peace.

Head Harbor was also the home of the Wilson family, who were deeply involved academically in the economics and science of the lobster fishing industry. Son Carl, who helped us out trapping the Boom Beach grid in 1993, has gone on to become the Director of Research for the Maine Department of Marine Resources.

In the early years of our research project, from 1990 to 1994, Isle au Haut was a close match for Monhegan in the rate of adult ticks flagged in the fall (7 to 15 per person-hour) and the percentage that were infected with the Lyme spirochete (15 percent to 34 percent). The ticks and the disease were just getting established then. For comparison, in 1994 at our study site in Wells we flagged over a hundred adults an hour, 57 percent infected. We continued to flag Isle au

Haut each fall through 1999. In the later years differences between Isle au Haut and Monhegan would be dramatic.

CHAPTER 3.

MATINICUS

There is another island even further offshore than Monhegan. Matinicus is 15 miles from the mainland in outer Penobscot Bay. There have never been deer on the island in anyone's memory. Ken Crowell (see Chapter 2, "Isle au Haut") and his son Tom had visited to study mice in 1989. They set out 112 mouse traps and captured 12 deer mice but found no ticks on any of them. Tom mentioned that there were muskrats and lots of cats on the island. Later, by phone, the postmaster confirmed the feral cats, claimed they'd cleaned out the raccoons and rabbits, but then said there were plenty of rats, even mentioning a few properties where there was a good chance of finding them. So, with no deer and a variety of small mammals, if a lack of established deer ticks on the islands could be corroborated, we'd have further evidence for the preponderant role of white-tailed deer in the deer tick life cycle.

Matinicus Harbor

Our first trip out was in mid-October, 1992, two of us and a friend who came along to help with the flagging. Getting out to Matinicus could increase the heart rate. The ferry from Rockland was too infrequent, took too long, and its schedule was weather-dependent, so we made the jump in a four-seat Cessna from the Owls Head airport. Fleeting by a thousand feet below, the islands of the Muscle Ridge Channel and West Penobscot Bay unfolded a magnificent panorama, but 12 minutes later came the windy landing on the north end of the island, fast, wobbly, and on an uphill grass strip cut perilously close to spruce woods on each side.

We were a little uncomfortable about how we mainlanders might be accepted on the island. This community of offshore fishermen was a good example of the relationship between isolation and independence. Autonomy reigned on the waterfront where dominance was based on the length of one's island pedigree, some going back over six generations. Bitter lobster wars had been fought, shots had been fired, and there was no constabulary. But that was one part of the community. As we were about to be left at the deserted airfield, a fellow passenger on the plane generously offered us a ride with his English Setter to Tuckanuck Lodge, the only accommodation on the island and a mile away.

Tuckanuck Lodge on Matinicus Island

The proprietor, Bill Hoadly, greeted us. Bill liked to talk, continuously and unbridled. It was difficult to break off, and tricky for us three to coordinate an escape to go flagging.

But we did, and in a day and a half, after a combined 18 hours of flagging covering all parts of the island, we found not a single tick, nor were there any on the five mice we were able to live-trap.

Our next trip out came in 2003. That time, we had another white-knuckle, crosswind, downhill landing in the Cessna. To get some local knowledge, we made our way to Paul and Eva Murray's house. Paul ran the island electric service and was tax collector. Eva was a freelance writer with frequent, thoughtful, articles in *The Working Waterfront* and other coastal publications. Both were very accommodating, pointing out trails we had missed. Later, on one of these, Chuck found the only deer tick during our nine cumulative hours of flagging.

At one point the trail popped out at Tuckanuck Lodge. Bill was there, and the conversation started right where we left off 11 years before. While chatting, we recovered two deer ticks from Mandy, his Golden Retriever. As on Monhegan, deer ticks were here, but precious few could be flagged. One more trip was made the next year. This time no ticks were found during 15 flagging hours, nor were there any on the three dogs who took up with us as we plodded along. Surely they had not become established on this deer-free but mouse- and rat-inhabited offshore outpost.

CHAPTER 4.

BACK TO MONHEGAN

The year that Katie Boegel sent the Plantation's request for a deer reduction to IF&W, 1996, was our seventh on the island. During that period the rate of adult deer ticks collected per hour each fall had risen from 8 to 17 and now a third were infected. But that year there were few larvae or nymphs on the 24 rats we captured, very possibly because the adults that would have created them were killed by the ivermectin-laced blood they imbibed from a treated deer in the two previous years. Despite this glimmer of success, many of the islanders were getting impatient to get on with a deer cull.

The number of Lyme cases was increasing as well, and deer had overrun the village, marauding gardens and stripping the plantings carefully put out by Kathie Iannicelli, the island's landscaper. Outside the village, they had heavily browsed edible vegetation, leaving warrens of impenetrable Japanese barberry. Our most productive, but most prickly, flagging (forty ticks per hour) had been in the barberry patch below the water tanks on Lighthouse Hill.

By 1996, 17 definite and nine probable Monhegan cases of Lyme disease had been reported to the Maine Bureau of Health since the first one in 1989. To estimate how many non-resident cases there might have been during that period, we developed a questionnaire for that much larger, potentially exposed population. In addition to all property owners and members of Monhegan Associates, we sent it to any person who had spent at least a week on the island between 1989 and 1995. Thanks to dedicated searches by the proprietors of seasonal lodging and of the island's two hotels, we assembled a list of 500 non-resident people. Our survey asked how long they had been on the island and if they'd contracted Lyme disease, then posed questions that enabled us to assure that they fit the case definition. Amazingly, three-quarters of the addressees took the time to fill out and return the questionnaire, adding 18 more cases to the list. That year there were six new cases among the 70 year-round residents, an alarming incidence. It was time to act.

In extensive reviews of the problem, IF&W biologists wisely concluded that the most efficient and least complex way to remove the deer on Monhegan would be to involve a professional marksman. They chose Anthony DeNicola, PhD, to do the job.

Tony had just completed his doctorate at Purdue based on deer population control. On that basis, and from seven previous years of wildlife control experience, he had just established a non-profit company he named White Buffalo. Its mission was "dedicated to the conservation of native species and ecosystems" using a "unique approach" involving "wildlife management alternatives in non-traditional settings". Tony's method was to dispatch deer over carefully selected bait stations at night using .223 caliber matched-grade rifles, special ammunition, night vision scopes, and silencers. His target was the center of each deer's brain. His team, including IF&W personnel and selected islanders, went to work in early April, 1997. Peter Boehmer provided the following account in that month's issue of *The New Monhegan Press:*

> Fifty-two deer were taken from April 2nd through April 6th, 1997. In a conversation with Luke Church (who has been working with Maine Medical Center in their effort to study and reduce the causes of Lyme disease on the island) it was stated that the State biologists know of at least ten deer still left on the island, though the most likely number is 17, with an outside possibility of 30 remaining. The deer were consistently underweight with no fat. The bone marrow on some indicated permanent damage from nutrition lack. Deer that should have weighed over 200 pounds (by the size of the animal) were in the 140 lb. range according to the cutters. The planning of this effort was without flaw. The shooting took place from dusk to about 2 a.m. The team was discrete, low key. While most of the island residents were pleased that something was being done to take control of the risk of Lyme disease, few, if any, were pleased at the thought of killing the deer, and some were more than upset by the process. The team consisted of two sharpshooters, three State biologists, two butchers (cutters), an assistant cutter and boatman, plus Luke. The meat was cut and prepared for taking inshore to be turned over to the State program providing this form of food to the needy. Scientific information will be gathered as a result of the Island project that will be helpful to deer management everywhere.

Gerry Lavigne reported that of the 52 deer harvested, only one realized that it had been shot, requiring the 53rd round. After analyzing biologic data, the biologists estimated that around 25 deer remained on the island.

A month after Tony's marksmen left, we went out with Gerry to review the program with the islanders. Fifteen came to the meeting. There were some basic misconceptions. Some thought that what had been done would eliminate deer ticks, and others, realizing it wouldn't, wanted immediate elimination. It was left that Gerry would write a letter estimating the number of deer remaining and reiterating the State's offer to do whatever the Plantation wished.

Chuck Lubelczyk, Rob Smith, Pete Rand, Eleanor Lacombe and Gerry Lavigne heading out to Monhegan on the Laura B.

Before leaving, we had one more task. We had become involved in a Phase 3 trial of a new Lyme vaccine manufactured by SmithKline, and two islanders had volunteered to have their blood drawn and be vaccinated. That turned out to be a last-minute affair consummated on the pier with the help of Luke Church as the *Laura B.* was churning to leave.

During this visit, Chuck trapped a rat near White Head with two deer tick nymphs on it. Its blood, spleen, and other organs plus the nymphs were sent that night to Sam Telford to help his search for the agent of a newly recognized tick-borne disease called human granulocytic ehrlichiosis (now anaplasmosis). In 1997, three of eight island dogs tested positive for exposure to this emerging tick-borne pathogen.

Gerry's letter to Chris Smith, then First Assessor of the Plantation, was five pages single-spaced. He pointed out that if they were not controlled, the remaining 25 or so deer could bring the population back to 100 per square mile in five years. Because fewer deer are harder to hunt, an annual program to bring and maintain a level of around five deer per square mile would require a lot of time and money, and require it indefinitely. Removing all the deer would be expensive, but because the island is isolated, the expense would be for one time. He outlined a plan to bring and maintain the herd to 12 deer per square mile with island hunters, but even that would involve establishing Monhegan's first tagging station, requiring duly licensed hunters passing a hunter safety course, and season dates and bag limits to be obeyed. His letter went unanswered.

In October 1997. we went back out for our annual adult tick flagging and serosurvey of the year-round islanders. Gerry went with us to have a meeting called by Assessor Mary Beth Dolan to discuss deer control options again. From a letter he sent to Chris Smith two days later:

> In discussion on the evening of the 23rd with Mary Beth, Rick, and a few other townspeople, it became evident that Monhegan Town Government has not yet come to a decision as to how deer should be managed (and/or studied). The only consensus we reached was that you did not want deer numbers to return to their 1996 levels. It is also my understanding that you will call a special meeting to vote on the proposal for MDIFW to hold the herd at 12 wintering deer, while monitoring tick and deer populations for the next five years. I was also informed that you may again vote in March on a warrant to completely remove deer from Monhegan.

That special Town Meeting took place on March 24, 1998. The vote to remove the remaining deer was 31 yes, 23 no. Lyme disease was not the only issue driving the votes for extirpation.

> "We had a problem and we corrected it, [Dave] Boegel said. "None of the conditions that merited killing the deer the first time exist here now. I think the gardeners voted me down." Winnie Murdock said the specter of Lyme disease haunting the island was too much..."
>
> *Manchester Union Leader,* April 3rd, 1998

With far fewer and more edgy deer on the island, the next cull took longer. In six days in late April 1998, Tony's crew removed 21 of the remaining herd, previously estimated to be around 25. Hair and entrails found in the woods gave evidence that two deer had been shot the day before and the day of the arrival of the sharpshooters. Some had tried to disrupt the hunt by seeking an injunction through the Plantation. That didn't work. But in the week before the cull, illegal bait stations were set up in secret places to lure deer from the cull sites, and guns were fired off to spook the deer.

Afterwards, Gerry estimated that eight to eleven deer were still there. He also determined that prior to the 1996 cull there were 105 to 108 deer on the island, or 127 to 130 per square mile. White Buffalo completed the removal during the following winter. The last deer was taken in March of 1999. Now, for the first time anywhere, we had the unique opportunity to observe the effect of complete and permanent deer removal on the abundance and infection prevalence of questing deer ticks.

But there's a paradox. Since the deer tick's life cycle is two years long, it will take that long before questing adults disappear. In the meantime, with no deer to feed on, the ticks that would have fed on them will now add to those questing for a host. In addition, we found that around half of all feeding nymphs do so on deer, which are not Lyme reservoir hosts. In fact, they kill spirochetes delivered by infected feeding ticks. Without deer, the nymphs will now feed on highly infected and transfer-competent rats, which will greatly increase the percentage of molted adults that are infected. For one or two years after the last deer was taken, then, we should expect to flag more, and more infected, adults in the fall until the cycle collapsed. Our annual trips continued.

The ride out the next fall was too exciting. From the trip record of October 26-28:

> Mary and I ran into Sam [Telford] at the muffin shop in Thomaston around 9:30. When we arrived at Port Clyde a half hour later we were met by Eleanor and Rob who announced that the *Laura B.* was canceled because a stiff southwest wind made things too dicey at the Monhegan pier. We arranged to call back at 1:00 to see if one of the several private charters would give us a ride, then drove to Rob's place in Cushing to eat our Italian sandwiches and catch up with what Sam was doing. The call back at 1:00 was that Robert Bracey of Monhegan was bringing his boat to the float and would take us out. Mary Beth Dolan wanted to go with us. Robert Bracey was a very pleasant young fishermen whose boat was a thirty-four foot Young Brothers lobster boat, which also served as a pilot boat for tankers plying Penobscot Bay. It was breezy but relatively flat out to Allen Island. Everyone crowded into the pilot house where there was precious little to hang on to. Robert had removed all the electronics except the radar so he could paint the interior, so the overhead bristled with empty clamps that threatened scalping when the boat began to pitch into the seas once out of the lee provided by of Allen Island. Beyond Old Man Ledge the seas built to ten to fifteen feet and the pilot house became a dice cup as the boat fell off the crests and plowed into the next mountain of water. I lost my grip while reaching for a camera and bounced off the plywood bulkhead on the starboard side of the pilot house, shattering it – then clutched the instrument panel box the rest of the way with open water rushing alongside. The check I gave Robert as his boat danced alongside the Monhegan pier was so waterlogged, I'm surprised that it could be processed.

Holding on for dear life

The high numbers of adult ticks and the proportion that were infected this year was telling. Just prior to the cull, we collected 17 ticks per hour, 31 percent of which were infected. Now we collected 28 ticks per hour, 74 percent of which were infected. This dramatic upward shift was testimony to the numbers of ticks the deer had been hosting and disinfecting previously, as well as to the extraordinary competence of the rats to transmit Lyme spirochetes to the larvae and nymphs that fed on them. But this year was the peak.

In the following October we shared a calm, warm, trip out on the *Laura B.* with a deck load of firewood. It was reassuring to get friendly greetings on the pier from several of the island's principals. The change in vegetation over one short year was striking. The underbrush had grown in thickly and places we used to flag had become impenetrable with native species. Although ticks were still plentiful in some areas, in others they were hard to find; island-wide the count had dropped back to 18 ticks per hour, 60 percent of which were infected. Chuck brought out live traps to capture both rats and muskrats. The muskrats were not fooled into the traps, but he captured 15 rats. We were overjoyed when no deer tick larvae or nymphs were found on any of them, sure evidence that there had been no deer-fed adults to produce them in the last two years.

As much as we looked forward to returning year after year to track the tick's decline, we also anticipated these trips as retreats, opportunities to extract wisdom from professional colleagues who might join us, to dream up new ideas for grant applications, to indulge in critiquing academic competitors, all loosened by a dram or two of Scotland's iconic beverage as the evening progressed. On one occasion, hoping to stimulate the group's ingenuity, a member who couldn't make the trip taped a dozen ecological hypotheses needing research down the side of a contributed amber bottle. Brilliant solutions were reported, but somehow never recorded.

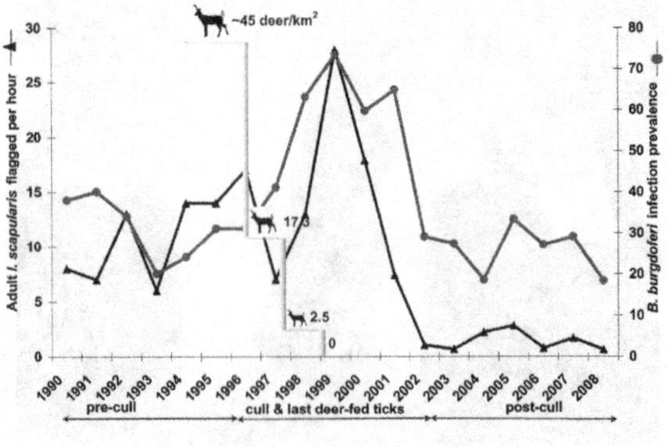

The black line indicates the sharp rise in fall-flagged adult ticks following the deer cull. The red line shows the rise in percent infected, as only rat-fed nymphs molt to adults.

We knew that cats and dogs were feeding adult ticks. From a survey we conducted in 2007, there were 15 cats and 8 dogs on Monhegan, but compared to the 113 deer that had roamed the island previously, these pets played a minimal role in the tick's life cycle. Whatever their contribution, there were no larvae or nymphs on rats to show for it. The ticks had virtually vanished, but why were around a third of the few we could flag still infected? Birds?

Rob Smith, June Ficker, and Chuck Lubelczyk processing mist-netted songbirds on the deck of Tribler Cottage on Monhegan.

From the beginning, the role that birds played in introducing, establishing, and spreading deer ticks and their pathogens intrigued us. In the early 1990's, collaborating with bird-banders at the Shoals Marine Laboratory on Appledore Island off Kittery, we found that about 2 percent of songbirds captured during their spring northward migration hosted one or more deer ticks nymphs. Not many, but in view of the countless flocks that arrive each year, this was a component to be considered in the introduction of tick-borne diseases.

In fact, by analyzing data from Appledore birds and the post-cull Monhegan flagging results, our statistician Susan Elias was able to show that the number of adult ticks we were finding on the island after the life cycle had been severed could be explained entirely by nymphs introduced by songbirds during spring migrations. In an earlier study we found up to 40 percent of nymphs removed from nesting songbirds in mid-summer were infected, so it was equally likely that the approximately 30 percent infection rate we found in Monhegan ticks after the deer were gone came from nymphs deposited by visiting songbirds.

CHAPTER 5.

THE BOTTOM LINE

The bottom line is that when the residents of Monhegan allowed the complete and permanent removal of white-tailed deer, they virtually eliminated the risk of contracting Lyme or other tick-borne emerging diseases on their island. From 1989 through 1997, 27 cases of confirmed or probable cases of Lyme disease were reported from the island, eight alone in 1996. The last possible transmission of a tick-borne disease from a tick derived from a Monhegan deer would have been in the spring of 1998. From that date forward, now 18 years, only two cases of Lyme disease, one in 2005 and one in 2011, have been reported from Monhegan.

In conclusion, this project would not have taken place were it not for a remarkably fortuitous cluster of events. Mark Wilson stopping by Monhegan asking about ticks. Martha finding and sending them to be identified. Rob Smith, a trainee of Andy Spielman, returning to Maine to find a lab in need of a new direction. An isolated island with ticks but no mice. What started out as a discrete research project evolved into a decade-long series of exploratory endeavors driven by an increasing threat to a community's health and livelihood, but constrained by its majority's resistance – which itself evolved. The actors were an eclectic group of field scientists, community leaders, conservationists, state biologists, federal animal control officers, and many individuals, including the avidly supportive and the quietly contrary with independence, dedication, wit, bureaucracy, diplomacy, trust, and a bit of eccentricity and subterfuge thrown in.

As a scientific effort, the Monhegan project contributed substantial new knowledge about deer ticks and what supports them. This is valuable information for epidemiologists and public health professionals far beyond the island:

- Where there are no mice, sub-adult deer ticks will be well fed and infected with the Lyme pathogen by Norway rats (Smith, RP Jr., et al., 1993). (Note: Please see the next section for full bibliographic references.)

- Deer mice, like white-footed mice, are very competent at feeding and infecting deer ticks (Rand, PW, et al., 1993).

- If deer consume enough ivermectin-treated corn, ticks feeding on them will die (Rand, PW, et al., 2000).

- The bite *of Ixodes muris,* the mouse tick, may be associated with a severe inflammation at the bite site in domestic animals (Lacombe, E, et al., 1990).

- When overabundant deer are completely and permanently removed from an isolated, Lyme-endemic community , the number of questing adult ticks and their infection prevalence will increase sharply until the pre-cull two-year life cycle ends, after which the vector ticks will virtually disappear and Lyme disease will become rare (Rand, PW, et al., 2004).

- The approximately 30 percent infection prevalence among the few adult ticks found post-cull is completely explained by the introduction of infected nymphs by migrating songbirds (Elias, SP, et al., 2011).

But these pages have really been about the art of field research, not the product, a subject that's rarely visible in the stilted texts of scientific publications. This project would have gone nowhere if the researchers had not first gained the trust of the islanders and our off-island collaborators. We had to understand and acknowledge sometimes strongly disparate viewpoints. Everyone needed a lot of patience to have conversations, spread information, and do their work. The project's success also depended on people like Luke Church and Faryl Henderson, who went way beyond the call of duty, and Tina Tulley, who bedded us down and entrusted us with her chancy vehicles. Character is another player, and so is humor. We researchers learned a lot more from Monhegan than we bargained for.

ACKNOWLEDGEMENTS

The writer has simply been the scribe for his team, for Rob Smith and Eleanor Lacombe from the very beginning, for Chuck Lubelczyk who cut his teeth on this project in 1995, for Mary Holman who was with us through the 1990's, for Susan Elias who has been with us since, and for our field crew in the early days, Marcus Russell and Steve Rich. On Monhegan, Luke Church's efforts were critical throughout, especially during the ivermectin project. Faryl Henderson did all the on-island organizing; Jack Bell provided his store for blood drawing; Peter and Ricki (Raquel) Boehmer, through their *New* Monhegan *Press* were our voice. We also greatly appreciated our hosts at Tribler Cottage, Martha Yandle, Richard Farrell, and Tanya, who was our feline source of countless adult deer ticks. Doug Odum's deployment and monitoring of 140 bait tubes surely deserves recognition. On Isle au Haut Tina Tulley, her accommodations and her vehicles, bless them, were absolutely critical. We thank Greg Runge for the deer hides he provided, and the unknown driver who brought them down to Portland. We much appreciate the cabin Ted Ames unknowingly provided and for his forgiveness for its invasion. Ken Crowell supplied a vital area on our learning curve of both small mammal handling and relict island theory, and we much appreciate Carl Wilson's flagging for us on Boom Beach.

Very particular recognition has to be given to Gerry Lavigne and his colleagues at the Maine Department of Inland Fisheries and Wildlife. Gerry guided us and the Monhegan islanders step by step through the several processes involved in treating, then culling, white-tailed deer. His contributions to meetings with island residents, the Plantation's Assessors, and ourselves, and the liaison he provided with his administrators, were critical. They were also delivered with authority and a twinkle.

ACKNOWLEDGEMENTS

TICKS IN MAINE

The Lyme & Vector-borne Disease Laboratory (VDL) is a component of the Maine Medical Center Research Institute in Scarborough, Maine. It was founded as the Lyme Disease Research Laboratory in 1988, in response to the early invasion into the state by *Ixodes scapularis/dammini*, the vector tick of the disease. Our initial mission was to study the complex ecology involved in the tick's spread and establishment in order to explore new strategies for its control, as well as to share the results of our research with public heath professionals and the general public. With the introduction of West Nile virus in 1999, our role expanded to include mosquito species that can, and are now, transmitting West Nile and Eastern equine encephalitis viruses in Maine. We collaborate closely with Maine CDC, other state agencies, and with colleagues in other major research laboratories nationwide. We invite you to view our two websites at www.mmcri.org/lyme. Our address is MMCRI, 81 Research Drive, Scarborough, ME 04074, and we can be contacted at (207) 396-8243 or ticklab@mmc.org.

Over the years, our work has allowed us to gather useful photos of ticks at various stages of their life cycle. We present some of them here to help people identify ticks, both nymphs and adults they may encounter.

Adult female deer tick (left). Adult male deer tick (center). Engorged female deer tick (right). Photos by VDL/MMCRI.

On the left is an adult female deer tick which has a peak season in late fall and another in the spring for those who did not obtain a blood meal in the fall. This tick transmits Lyme and other pathogens. In the center is an adult male whose only purpose is to mate, not feed. The left photo demonstrates how engorged a fed female can become.

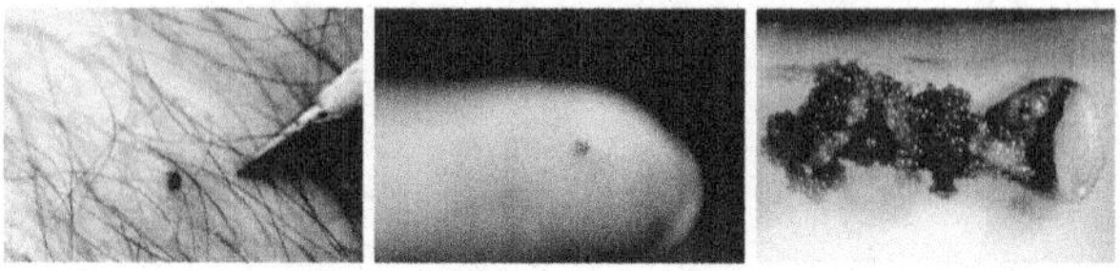

Feeding deer tick nymph (left). Deer tick nymph (center). Female deer tick depositing eggs (right). Photos by VDL/MMCRI.

Nymphs are responsible, by far, for the majority of tick-borne illnesses. They appear in late May and reach a peak in June and July, when the risk of getting Lyme disease is greatest. At far right is a female deer tick depositing eggs, as many as 3000. This happens in late May and early June.

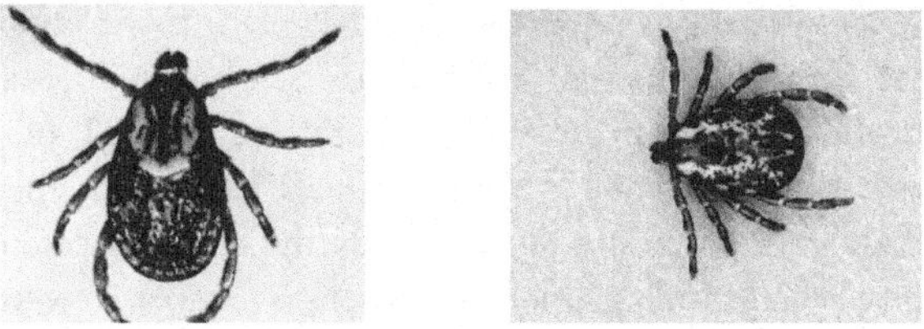

Female dog tick (left). Male dog tick (right). Photos by VDL/MMCRI.

These are dog ticks (*Dermacentor variabilis*), female on the left, male on the right. They are larger than the adult deer ticks. Note the white markings in the female and the striations on the male. They are very common in tall grass from April through July, are aggressive biters, but do not carry Lyme disease.

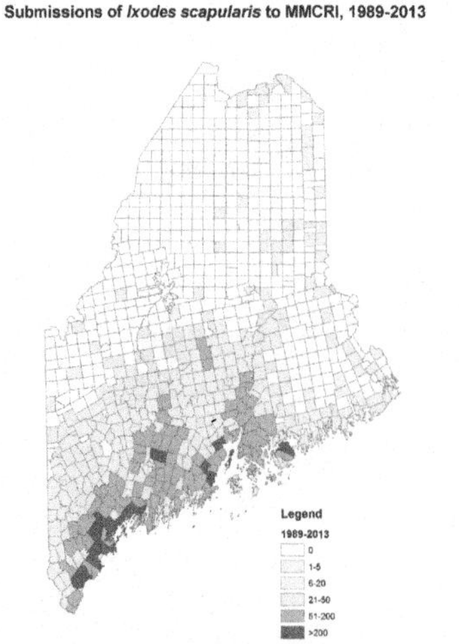

Chart by VDL/MMCRI and used with permission.

Beginning in 1988 through 2010, our lab offered free identification of ticks sent to us by the public, physicians, veterinarians, hospitals, and clinics. We examined more than 30,000 ticks, representing 14 different species. Shown in the map below are deer tick submissions, as the ticks first invaded the south coast, then advanced north and east, up river valleys, and into northern counties, where because of low deer populations, they are still uncommon.

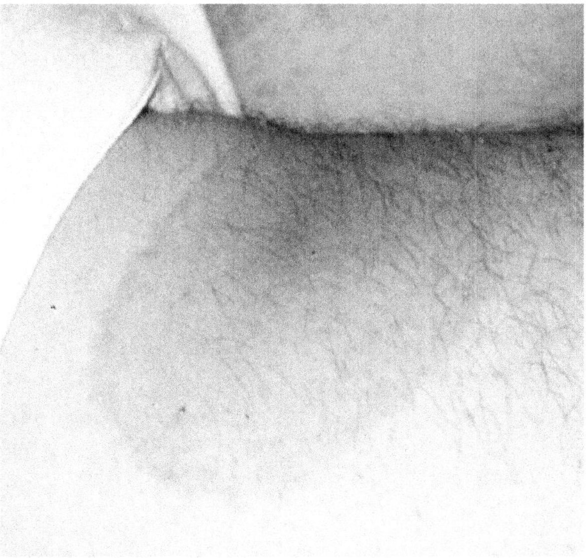

Erythema migrans, the classic rash of Lyme disease. Photos by VDL/MMCRI.

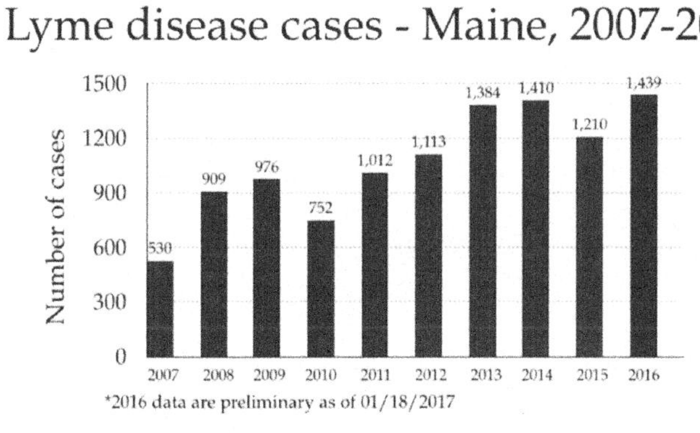

Lyme Disease prevalence in Maine, 2007-2016. Because of under-reporting, it is estimated that case numbers should be multiplied by about ten. (Source: Maine CDC Report to Maine Legislature on Lyme Disease, February 2017).

SELECTED PUBLICATIONS OF THE LYME & VECTOR-BORNE DISEASE LABORATORY AT MMCRI

Smith, RP Jr., et al., 1990. Potential for Lyme disease in Maine: Deer survey of distribution of *Ixodes dammini,* the tick vector. *American Journal of Public Health,* 80:333-335.

Rand, PW, et al., 1991. Canine seroprevalance and the distribution of *Ixodes dammini* in an area of emerging Lyme disease. *American Journal of Public Health,* 81:1331-1334.

Smith, RP Jr., et al., 1992. Diversity of tick species biting humans in an emerging area for Lyme disease. *American Journal of Public Health,* 82:66-69.

Lacombe, EH, et al., 1993. Disparity of *Borrelia burgdorferi* infection rates of adult *Ixodes dammini* on deer and vegetation. *Journal of Infectious Diseases,* 167:1236-1238.

Rand, PW, et al., 1993. Competence of *Peromyscus maniculatus* (Rodentia: Cricetidae) as a reservoir host for *Borrelia burgdorferi* (Spirochaetares: Spirochaetaceae) in the wild. *Journal of Medical Entomology,* 30:614-618.

Smith, RP Jr., et al., 1993. Norway rats as reservoir hosts for Lyme disease spirochetes on Monhegan Island, Maine. *Journal of Infectious Diseases,* 168:687-691.

Rand, PW, et al., 1996. Low seroprevalence of human Lyme disease near a focus of high entomologic risk. American *Journal of Tropical Medicine and Hygiene,* 55:160-164.

Smith, RP Jr., et al., 1996. Role of bird migration in the long-distance dispersal of *Ixodes dammini,* the vector of Lyme disease. *Journal of Infectious Diseases,* 174:221-224.

Kierans, JE, and EH **Lacombe,** 1998. First records of *Amblyomma americanum, Ixodes* (Ixodes) *dentatus,* and *Ixodes (Ceratixodes) uriae* (Acari: Ixodidae) from Maine. *Journal of Parasitology,* 84 (3):629-631.

Rand, PW, et al., 1998. Participation of birds (Aves) in the emergence of Lyme disease in southern Maine. *Journal of Medical Entomology,* 35:270-276.

Lacombe, EH, et al., 1999. Severe reaction in domestic animals following the bite of *Ixodes muris* (Acari: Ixodidae) ticks. *Journal of Medical Entomology*, 36(3):227-232.

Rand, PW, et al., 2000. Attempt to control ticks (Acari: *Ixodidae*) on deer on an isolated island using Ivermectin-treated corn. *Journal of Medical Entomology*, 37(1): 126-133.

Smith, RP Jr., et al., 2001. Outbreak of Powassan Encephalitis-Maine and Vermont 1999-2001. *Mortality and Morbidity Weekly Report*, 50(35):761-764.

Rand, PW, et al., 2003. Deer density and the abundance of *Ixodes scapularis* (Acari: Ixodidae). *Journal of Medical Entomology*, 40(2):179-184.

Smith, RP Jr., 2003. Lyme Borreliosis: Biology, Epidemiology, and Control (Book Review). *Clinical Infectious Diseases*, 37:1724-6.

Holman, MS, et al., 2004. *Anaplasma phagocytophilum, Babesia microti,* and *Borrelia burgdorferi* in *Ixodes scapularis*, southern coastal Maine. *Emerging Infectious Diseases*, 10:744-746.

Rand, PW, et al., 2004. Thermal accumulation and the early development of *Ixodes scapularis. Journal of Vector Ecology*, 29:164-176.

Rand, PW, et al., 2004. Abundance of *Ixodes scapularis* (Acari: Ixodidae) after the complete removal of deer from an isolated offshore island, endemic for Lyme disease. *Journal of Medical Entomology*, 41:779-784.

Lubelczyk, CB, et al., 2004. Habitat associations if *Ixodes scapularis* (Acari: Ixodidae) in Maine. Environ. *Entomology*, 33(4):900-906.

Stone, E, et al., 2005. Antibody testing and Lyme disease risk. *Emerging Infectious Diseases*, 11(5):722-724.

Smith, RP Jr., 2005. Current diagnosis and treatment of Lyme disease. *Comprehensive Therapy*, 31(4):284-290.

Holman, MS, et al.: 2006. A checklist of mosquitoes of Maine with new state records. *Journal of American Mosquito Control Association*, 22:327-329.

Smith, RP Jr., et al., 2006. *Borrelia garinii* in seabird ticks (*Ixodes uriae*), Atlantic Coast, North America. *Emerging Infectious Diseases*, 12:1909-1912.

Elias, SP, et al., 2006. Deer browse resistant exotic-invasive understory: an indicator of elevated human risk of exposure to *Ixodes scapularis* (Acari: Ixodidae) in southern coastal Maine woodlands. *Journal of Medical Entomology*, 43:1142-1152.

Lubelczyk, CB, et al., 2007. First U. S. record of the hard tick *Ixodes* (*Pholeoixodes*) *gregsoni. Journal of Parasitology*, 93: 718-719.

Rand, PW, et al., 2007. Passive surveillance in Maine, an area emergent for tick-borne diseases. *Journal of Medical Entomology*, 44(6):1118-1129.

Cahill, B, et al., 2009. Anaplasmosis and ehrlichiosis – Maine, 2008. *Mortality and Morbidity Weekly Report*, 58(37): 1033-1036.

Lubelczyk, CB, et al., Tick (Acari:Ixodidae) infestation at two rural, seasonal camps in Maine and Vermont. 2010. *Journal of Parasitology*, 96:442-443.

Rand, PW, et al., 2010. Trial of a minimal-risk botanical compound to control the vector tick of Lyme disease. *Journal of Medical Entomology*, 47:695-698.

Rand, PW, et al., 2011. Multitarget test for emerging Lyme disease and anaplasmosis in a survey of dogs, Maine, USA. *Emerging Infectious Diseases*, 17:899-902.

Elias, SP, et al., 2011. Density of *Ixodes scapularis* ticks on Monhegan Island after complete deer removal: A question of avian importation? *Journal of Vector Ecology*, 36:11-23.

Mathers, A, et al, 2011. Strain diversity of *Borrelia burgdorferi* in ticks dispersed in North America by migratory birds. *Journal of Vector Ecology*, 36:24-29.

MacQueen, DM, et al., 2012. Genotypic diversity of an emergent population of *Borrelia burgdorferi* at a coastal island recently colonized by *Ixodes scapularis*. *Journal of Vector Borne and Zoonotic Diseases*, 12:456-461.

Smith, RP Jr., 2011. The vectors of Lyme disease, Chapter 1. In *Lyme disease: an evidence-based approach*. (J. Halperin, ed.), CABI Press, Oxford.

Lubelczyk, CB, et al., 2013. An epizootic of Eastern equine encephalitis, Maine, USA in 2009: Outbreak description and entomologic studies. *American Journal of Tropical Medicine and Hygiene*, 88:95-102.

Elias, SP, et al., 2013. Effect of a botanical acaricide on *Ixodes scapularis* (Acari:Ixodidae) and non-target arthropods. *Journal of Medical Entomology*, 50:126-136.

Lubelczyk, CB, et al., 2014. Detection of Eastern encephalitis virus antibodies in moose (*Alces americana*, Maine, 2010. *Journal of Vector Borne and Zoonotic Diseases*, 14:77-81.

Lubelczyk, CB, et al., 2014. Parasitism of mustelids by ixodid ticks (Acari:Ixodidae). Maine and New Hampshire, USA. *Ticks and Tick-borne Diseases*, 5:432-435.

Smith, RP Jr., et al., 2014. Human babesiosis, Maine, USA, 1995-2011. *Emerging Infectious Diseases*, 20:1727-1730.

PHOTO PERMISSIONS

Unless otherwise marked, all photos were taken by the author, Dr. Peter Rand, and are used with his permission.

Made in the USA
Monee, IL
04 November 2026

67377604R00039